THE FUTURE OF INTELLIGENT AUTOMATION

How AI and RPA can improve your business processes!

BY:

ELIJAH FALODE

Copyright © 2020 *Elijah Falode*

All rights to this book are reserved. No permission is given for any part of this book to be reproduced, transmitted in any form or means, electronic or mechanical, stored in a retrieval system, photocopied, recorded, scanned, or otherwise. Any of these actions requires the proper written permission of the publisher.

Contents

PREFACE 1
CHAPTER 1 1
A HISTORY OF INTELLIGENT AUTOMATION 1
   History of Automation 2
      Communication Technology and Intelligent Automation in History 4
      The Popular QUESTION frequently asked 5
CHAPTER 2 7
WHAT IS INTELLIGENT AUTOMATION? 7
      Once Again: Are we trying to Replace Humans? 9
      What is intelligent automation teaching us? 10
      The Vast Potential of Intelligent Automation 11
CHAPTER 3 14
ARTIFICIAL INTELLIGENCE VS. INTELLIGENT AUTOMATION 14
   What is Artificial Intelligence? 14
      All About AI 14
   The Four AI Approaches 16
   How is Artificial Intelligence Classified? 16
   The Importance of Intelligent Automations 18
   The Difference Between AI (Artificial Intelligence) and IA (Intelligent Automation) 21

A Complementation of Concepts and Functions 22

CHAPTER 4 24
AUTOMATION IN INDUSTRIES 24
    What is automation in the industry? 24
    Why Is Industrial Automation Today's Trend? 26
    Types of Industrial Automation 27
        Fixed Automation 27
        Programmable Automation 28
        Flexible Automation 29
    Industrial Automation for Quality and Flexibility in Production Process 29
    The Advantages of Industrial Automation 30
        Lower Operating Cost 30
        Increased Productivity 31
        High Quality 31
        Increased Accuracy in Information 31
        Improved Safety 32
    Disadvantages of Industrial Automation 32
        High Initial Cost 32
        Fear of Retrenchment 32
        Loss of Flexibility 32
CHAPTER 5 34
EXPLORING THE TRENDS 34
    What are the Trends in Intelligent Automation? 34
        Digital Transformation Trends 35
        Unattended and Attended Automation and AI 35

- Governance Gains Focus 36
- Creating Order from Chaos 36
- Chief Information Officers Become the Center of Focus 37
- Ground Breaking Performance 38
- Automation in Decision Making 39
- Test Automation 39
- Easy Development of Automations 40
- Keeping Pace with IT Operations 41

CHAPTER 6 43
ARE THERE PITFALLS? 43
- Overspending 44
- Gains which do not Commensurate Investments 45
- Allowing Artificiality to Reign over Natural Intelligence 46
- Trying to Automate Everything 47
- Neglecting Automation 48
- Replacing Humans Totally 48
- Thinking Automation Doesn't Need Modifications 49
- How to Avoid the Pitfalls of Automations 50
  - Get Everyone Onboard 51
  - Plan to Scale Automation from the Beginning 51
  - Establish Awareness 52
  - Knot the Ties of Information Technology and Automation 52

    Ensure the Tasks Need Automation 53

    100% Shouldn't be Given to Automation 54

CHAPTER 7 55

THE FUTURE OF INTELLIGENT AUTOMATION 55

    The Future Includes You 55

    The Place of Automation in the Future 56

    More on the Future 57

CONCLUSION 59

ABOUT THE AUTHOR 60

REFERENCES 61

PREFACE

Since the 1800s, automation has been assisting humans in taking up roles in different industries like finance, travels, healthcare, education, and different sectors of life. This invention has developed to what we now see as intelligent automation because of its major capacity to learn and adapt to novel situations through a process called deep learning. Today, intelligent automation is performing amazing roles and making life much easier for us.

Intelligent automation is deployed for tasks such as customer care, fraud identification, teaching, and facial recognition. Because it's making so much impact in today's world, we're tempted to seek the future of automation.

The future of automation is roped with the certainty of robotic software, not just being automated but intelligent. The future will have humans engaged only in major roles such as programming, creativity, and supervision because automation has become highly intelligent enough to finish up simple and complex tasks.

When you read the first chapter of this book, you'll understand how automation has evolved over many years to become what it is today---

an intelligent software. It is now about machines that can understand their environment, interact, work with humans and other machines, and even learn from personal experiences.

In the second chapter, I explained the concept of intelligent automation in the simplex and understandable way.

Chapter three clears the air between similar and complementary technological concepts; intelligent automation (IA) and artificial intelligence (AI). It gives you insight into how factors such as machine learning, natural language processing, predictive analysis, and robotics are combined in new technologies. An example of this complementary invention is driverless cars, which use numerous inventions such as **sensor inputs**, the **camera feeds**, historic archive, and **real-time machine feedback** to give a comfortable and **secured driving experience** that you've never had.

If you stop to read in chapter three, you may not capture a very peculiar aspect of intelligent automation, which you'll find in chapter four. It discusses the effect of automation in industry. You'll understand how automation helps in the customization of products and services, reduction of production or delivery cost, and improvement of efficiency. In marketing, for instance, offers are personalized according to clients' profiles and choices, while automation is resourceful in identifying and putting a stop to fraudulent transactions in credit card processing. Intelligent automation has become so useful in industries that you can hardly picture a bank without ATMs or a store without a POS. It is now an important aspect of most industries in today's world.

The trends of intelligent automation are highlighted in chapter five, while the pitfalls of intelligent automation are explored in chapter six. Some of the pitfalls of intelligent automation include overspending, allowing artificiality to reign over natural intelligence, trying to automate everything, and replacing humans outrightly, and some other adoption mistakes which every knowledge seeker on intelligent automation needs to read in chapter six.

In the last chapter, I recollect how intelligent automation will be the mainstream tool of the future. And there will be more advanced intelligent automated software in the future. These machines will possess a higher degree of adaptability features and an unprecedented level of productivity and value propositions.

This book is like the manual and scripture of intelligent automation because its content reveals everything we currently see of intelligent automation and what to expect in the future. It's an informative material for every individual, group, or company that desires to take the lead in its sector. If you are one of the few potential successful persons who want to have a share of the future before others begin to scramble for the crumbs, then follow me as I take you from the first to the last chapter of this automation scripture.

CHAPTER 1

A HISTORY OF INTELLIGENT AUTOMATION

In today's world, we see automation everywhere and in almost everything we do. Its impact is like a vibrant tune that plays from our workplaces down to our very own home. Look around you, look into every industry of today's century, and you'll see humans leaping innovation through automation. And you know what? It's ever-increasing. It never stops. It is a transition from one development of innovation to another, leaping with little or no bounds and making an impact along the way.

Now we have Siri, which is there to answer your command. It efficiently opens your playlist and gives you the song you have commanded it to play. And did you know that Google Assistant can offer you virtual support and schedule for you a seat at your favorite restaurant?

What about having an automated driver designed to pick you up from your home to your destination at your beck and call? Drones are now offering delivery services. The impact of intelligent automation is becoming innumerable. Before now, we only saw this on screens and not in real-life situations. Who could have thought the evolution of machines could happen this fast? As this elasticity of innovation dazzles our reality, we are drawn towards the question: when did intelligent automation begin?

History of Automation

So many centuries before the 21st, early man had to carry his hunting task through different weapons carved from stone. It was the Stone Age. Humans defended and catered for themselves through this invention. Then there was milestone achievement sometime around 4000BC. It was the manufacturing of the wheel, and it led to what we have today. The **locomotor** movement became easier because of this invention. Transportation became faster, cheaper, and exciting. During the first century of human existence (10-70 AD precisely), the Greek scholar and mathematician Heron of Alexandria invented what the world has recorded as the first automated door ever made. Through a mechanism made up of some ropes and pulley, it gave passage into the city. Let's now turn our history pages and peer through the 18th century.

The industrial revolution paved the way. It was a century of man's show of creativity. Man's natural endowment came to the limelight. Most facilities in that era were manufactured to run with water and steam. The start of the 20th century built on this development and there was the mass invention of production facilities which made use of electricity. The era was known as the Industrial Revolution 2.0. And you may be curious to know if there was a 3rd industrial revolution. Yes, there was the industrial revolution 3.0, which happened in the 1970s. It was the era that had a shift towards the wide usage of electronic devices and Information Technology. Could there have been an Industrial Revolution 4? Yes, there is, and it's of today's world. We call it "Industrial Revolution 4.0". It's based on the use of AI (Artificial Intelligence) and **cyberspace.** Not just this, automation technology is having a paradigmatic shift through creative inventions such as **Blockchain, Intelligent Chatbots**, and **Integrated** or **General AI.**

Communication Technology and Intelligent Automation in History

It's difficult to discuss the history and evolution of automation without highlighting that of communication technology because both developments took place simultaneously. Communication technology has advanced over a long period. We can trace it back to the time of cave paintings at about 40,000 BC to the 18th century when telegraphing was in vogue and down to today's world of internet communication. Communication technology has contributed a lot to our persona-socio relationships. It has inadvertently contributed to the easy access and information transfer between persons and groups, and it has resulted in what we now call the evolution of automation.

The pace at which the evolution of communication technology and automation takes place is as fast as our mind can imagine, and our hands can act. It's like an upsurge that is not meant to take a break. Every organization which leads in today's world is adapting quickly to the advent of this innovation. One of how these organizations are embracing automation is their usage of **RPA (Robotic Process Automation)**. This adoption supports the speculations that Artificial Intelligence, in the chart of technology, is the next big thing.

We can already see how **Big Data**, **Natural Language Processing**, and **Machine Learning** are utilized in areas such as health, finance, marketing, education, law, etc. The harmonization of the Natural Language Processing (NLP), ML – Machine Learning, and RPA – Robotic Process Automation has caused so many positive disruptions in the aspect of Intelligent Automation. To successfully benefit from this new technology, we have to be flexible for adaptability and innovation.

The Popular QUESTION frequently asked

Does intelligent automation have a negative impact?

In the last two chapters of this book, I gave an in-depth answer to this question. But to briefly answer it now, I will say that the answer is in two ways, Yes and No. Yes because when intelligent automation is compared with the innovations of industrial revolutions in time past, we see that these kinds of innovations usually make impacts and affect the generation they exist in.

However, blue-collar jobs are reducing in numbers every day, and this is because automation is replacing humans in different industries of our world. Most white-collar jobs in our present time are dropping off humans and recruiting intelligent automated devices, especially in jobs that do not need much creativity or the repetitive jobs.

I also said the answer could be no because as obsolete skills are laid off, those with skills that can operate the new technology are hired. The need for new information and skills has been increasing over the centuries. Humans had to learn to ride chariots. We left the typewriter to learn how to operate the computer, and we are here now. Learning is a constant process. So, as the century demands new skills, we won't say it's the fault of intelligent automation. No, it's not. It's the fault of the man who is not ready to learn the skills required of the intelligent automation age. This and more would be discussed in chapters six and seven of this book. So, ensure not to stop halfway. Read and enjoy to the last!

CHAPTER 2

WHAT IS INTELLIGENT AUTOMATION?

Artificial Intelligence and Intelligent Automation are the buzz words freaking out the technological sphere today.

Everything seems to revolve around them. And everybody seems interested in talking about these peculiar terms.

It makes our mind to be curious about the definitions.

Before now, the capacity of computers could not measure up to the demands of applying artificial intelligence to solve complex problems. Even as we relied on the use of this invention, we had to face other limitations. Take, for instance, driverless cars require trustable sensors to get whatsoever information the car will need to react to situations. Although modern technology has improved tremendously, there are many more milestone achievements that are attainable now, even if they weren't many years ago. So, what is intelligent automation?

The prospect and definition of intelligent automation become clear when we understand its background. It's the compounding of two technological concepts that have existed in history for long before now; **that's automation and artificial intelligence.**

Artificial intelligence, AI, is a relatively modern invention that has to do with machine learning, recognition of language, and vision, whereas automation has been in existence since the industrial revolution era. Both artificial intelligence and intelligent automation have been in technological advancements for a very long time. They have emerged to what we now have as **intelligent automation**, which is described as **a kind of automation that has the advantage of being artificially intelligent**.

Most technology-savvy persons know of RPA – Robotic Process Automation, and the advanced form of this is Intelligent Automation. It's a software that attempts to imitate the personality of a user via the use of web pages or enterprise application screens to search, assess, calculate, redefine, and place the data into fields of enterprise that are already existing and obeying the rules of business.

It has a transformative impact because of its cost reduction nature, deployment of high valued personnel, and branding of the product or service. All of these give the business a competitive advantage.

As a technology, intelligent automation is best for insurance purposes, especially those that demand labor: for example, policy administration, finance, business monitoring, etc. Large organizations that deal with repetitive tasks like giving information to so many systems need intelligent automation software.

Once Again: Are we trying to Replace Humans?

The straight answer to this is "no" because intelligence is not trying to replace humans with machines or software. Instead, it is helping to facilitate the efficiency in getting things done. It combines all the advantages of robots with that of human judgment. An automated device will not require some training, does not lose focus, and doesn't make errors if it is well programmed. It goes with the set pattern and doesn't deviate from this every time it does its duty. Exceptions or situations that are not in its programming sequence are flagged for human assessment.

However, nothing can accurately substitute human intelligence, especially when business circumstances demand critical judgment for a contextual course of action. Most times, RPA solutions are not just for bulky and regular tasks; they also leverage complex exceptions that need the evaluation of professional employees. Once they are sorted out and validated by these experts, they are sent back to the intelligent automation process. From this, we see the interdependence of humans and intelligent automation towards the goal of efficient and quality outcomes.

What is intelligent automation teaching us?

We can learn so much from intelligent automation. Right now, we are merely scratching the surface of its advancement. More evolutionary changes will unseat the present age of intelligent automation. In chapter seven, more insight into the future of this technology, intelligent automation, will be discussed.

Intelligent automation easily provides feedback information on activities it processes for an organization. Companies, both small and big, can use this data to understand and maximize their labor capacity, analyze the reasons behind every process exceptions through Pareto principle; establish where, when, and how to satisfy customers; and get insight into new business opportunities upon examining the profitable standpoints of market demands, economic factors, service, and product.

For every business and household, now is the age of intelligent automation. If you have a company, you need to deliberately invest more resources in acquiring intelligent automated devices because this cost-effective action will lead to an increase in productivity, error reduction, reduced scalability cost, and improve service quality to customers.

The Vast Potential of Intelligent Automation

"Why does intelligent automation seem so important? Are there special applications of intelligent automation?" These are questions asked almost everywhere. You, too, might have asked these same questions. Let's examine some potentials of intelligent automation.

Information Processing

One wonderful function of intelligent automation is its usefulness in processing vast amounts of data into information, making the data useable. Also, it can keep a record of and evaluate workflows, and this is because it's designed to be "intelligent."

You may be surprised if I told you that we do use intelligent automation in everything around us, and it ranges from cognitive computing and quality control, robotics and autonomous cars, efficiency, and business processes.

Big businesses are thoroughly searching for real opportunities in intelligent automation and making proper investments. Here are some practical illustrations that show the vast potentials of intelligent automation:

Google has bought over more than eight robotics startups; **automakers** are working tirelessly to design and produce autonomous cars. IBM is stashing so much money in **cognitive computing technology**. In case you are a small scale business owner or intend to start small, and you too want to join the intelligent automated trend, know that it's a technological advancement, anyone can stake their investments.

Processing of Situational and Textual Information
Because of our inherent intelligence, humans can analyze situations, read, and comprehend texts. Now, machines are doing the same thing---and they even do it more efficiently than what our human mind could naturally do.

Driverless cars are already being tested on major roads. These driverless cars make informed decisions through the situational data that has been inputted into it.

On a business level, so many businesses are using intelligent automation, yet many are not even aware of it. Let's say you have a business that runs on a website. Your website will be presenting information to your visitors based on their browsing history and pattern of purchase. This is the action of intelligent automation taking place.

More on how intelligent automation affects business and other usages will be presented in chapter four of this book.

CHAPTER 3

ARTIFICIAL INTELLIGENCE VS. INTELLIGENT AUTOMATION

What is Artificial Intelligence?

Every day, new technological breakthroughs are introduced to our world. And across different technological revolutions, we have arrived at a trend called artificial intelligence. You have probably heard of the word somewhere, but how well can you tell of the meaning? In order not to overwhelm you with much technological jargon, I have tried to use simple terms that are elementary to everyone. So, let's begin. What is artificial intelligence?

All About AI

Whenever the words artificial intelligence comes to mind, so many people tend to relate it with robots. This is because most movies we watch and novels we read weave their stories about human-like machines on Earth. But nothing could be further from the truth.

AI is a vast discipline of computer science that is focused on developing intelligent machines smart enough to carry out tasks that will normally require the intelligence of a human. It's a multidimensional discipline that cuts across other fields of science, and there are milestone improvements of the tech industry, especially as machine learning and deep learning keep advancing.

So, you can simply say that artificial intelligence refers to programmed machines simulating human intelligence to have the mindset and behavior of humans. These machines can be developed to have the capacity for learning and providing solutions to complex problems. In other words, the basic meaning of AI (Artificial Intelligence) is that machines can copy human intelligence to perform simple and complex tasks. Artificial intelligence is developed to make machines learn, reason, and perceive.

You can easily identify artificial intelligence through its rational and behavioral abilities, which are centered on achieving an objective.

 New advancements in technology create a redefinition of artificial intelligence. Before now, artificial intelligence included machines that calculate some basic functions or recognize texts through optimized recognition of character, but now these functions are seen as normal functions of a computer system.

The Four AI Approaches

According to Norvig and Russell, four different approaches appropriately describe AI [1]:

1. Thinking humanly
2. Thinking rationally
3. Acting humanly
4. Acting rationally

The first two are based on logical thinking, while the last two are focused on behavior. This approach defines artificial intelligence as rational agents who take up certain tasks for better productivity.

How is Artificial Intelligence Classified?

It's made up of two major categories: Narrow AI and Artificial General Intelligence.

Narrow AI: It's sometimes called "weak AI." It's limited in scope and imitation of human intelligence. A machine that is categorized as narrow AI is based on a specific task which it does excellently. All around you are evidence of narrow AI. It has always been evolving with a record-breaking impact on society.

Artificial General Intelligence (AGI): It's sometimes called "Strong AI." This is the type of artificial intelligence shown in movies like Star Trek. Artificial General Intelligence is a machine that is almost as identical as human intelligence because it can be applied to solve practically any problem.

When you see a robot performing human tasks or a digital computer, you should understand that artificial intelligence is at work. It's usually related to the process of developing a technology that can think logically, discover meaning, and learn from previous experiences. Since when the computer was found in the 1940s, we have seen that computers generally can be programmed to expertly perform different difficult tasks like playing chess and finding proofs of mathematical theorems.

As we see computers advancing in leaps, it has not been easy finding a program that can easily match humans' unique ability to be flexible according to a whole new level of context, which may be unprecedented. Nevertheless, some programs have been built to imitate humans in carrying out professional functions successfully. That's why you can see artificial intelligence making the diagnosis in medical treatment, optimizing computer search engines, and accurately recognizing voice and handwriting.

Through artificial intelligence, machines can acquire new knowledge from their experiences and environment, make concrete adjustments, and execute tasks that are meant for humans. You must have heard of self-driving cars, chess-playing computers, and other kinds of AI-enabled devices. They depend mostly on deep learning and natural language processing. These processes involve training computers towards achieving specific tasks through the processing of a large variety of data, understanding their patterns, and utilizing them for better future performance.

The Importance of Intelligent Automations

Why is artificial intelligence trending the technological sphere? What are its value prospects? If you have ever asked this question, read on if you want to clear them off your plate:

- AI is the automation of reoccurring learning, which charts discoveries through the deep learning of the prominent quantity of data introduced into it.
- Nonetheless, we should be careful not to mistake hardware-driven, robotic automation for AI; they are

not the same thing. AI does not automate tasks that can actually be done manually. It is, instead, used in performing time-consuming, high volume, digitalized tasks accurately without stress. This doesn't mean it works solely on its own. Occasionally, it relies on humans and makes inquiries on the processing of certain tasks.

- AI gives additional value to already existing products. Have it in mind that AI is often embedded within a device and not sold solely as an application. An already existing product will be developed through features or components of AI, just as Siri became an additional update to the latest Apple devices we have today. Technologies at home can be improved with automation, conversational platforms, bots, and a considerable amount of data.

- AI has made programming much simpler than it has always been. AI can adapt to algorithms that are progressively learned with experience so that both its inputted data and progressively learned patterns could perform the programming on their own. AI studies the regular patterns of algorithms and acquires new knowledge and skills to aid its future performances: the algorithm stands out like a predictor. In other words, AI is that chess which has learned by experience the best moves to make and what product should be recommended online. AI learns through the back-propagation technique. It's a technique that remodels its design based on training and fresh data.
- AI gives in-depth analysis through neural networks that have lots of

hidden layers. Before now, there wasn't much of a possibility in establishing a fraud detection system of five hidden layers.

Well, no one has to tell you that those days are gone by, and this is because of the advancement in the power of the computer. You will need to feed your models with a large bank of data. You may be interested to know why this is necessary: Deep learning models become more accurate by the quantity of data it is fed. Take, for example, Alexa, Google Photos, and Google Search are all possible because of deep learning; and as you use them more often, they tend to get more accurate.

The Difference Between AI (Artificial Intelligence) and IA (Intelligent Automation)

We have discussed so much about Artificial Intelligence, and now we will place it side by side with intelligent automation. We have seen that Artificial Intelligence (AI) has been trending recently, and it's really handy in slaying down complex problems in different fields such as health, gaming, economics, weather forecast, and so many more.

As introduced in chapters one and two of this book, we have also learned that automation is a modern-day innovation because it has traveled through different revolutions and came to limelight during the fourth Industrial Revolution. All over the world, we see companies adopting RPA – Robotic Process Automation to automate human tasks. You can see the job of clerks automated by using machines such as Automated Teller Machines (ATM) and point of sale (POS) devices. So what is the difference between Intelligent Automation and Artificial Intelligence?

A Complementation of Concepts and Functions

AI is limited if it's made to function in isolation. AI is a great invention for solving complex problems, but it's not the best for solving digital tasks that we face every day. For example, it's not quite easy to use artificial intelligence to tell if a PDF is an invoice or not. The need for complementation of both techniques is a transformative one.

However, you may still want to know that what differentiates artificial intelligence from intelligent automation is that artificial intelligence is designed to be autonomous workers. They can simulate or imitate human cognitive functions. In contrast, intelligent automation is based on developing more efficient workers, which can be digital or humans who are ready to embrace and work with intelligent technologies.

What you have just learned in this chapter is that intelligent automation is necessary technologies every organization must have because of its transformative impact. And, more importantly, organizations which adopt these two technologies early have proven themselves ready to claim the future of their field.

CHAPTER 4

AUTOMATION IN INDUSTRIES

Today, the growing competitiveness we see in different industries demands high quality and authentic products at a competitive rate. Several industries are considering new product designs and integrating new manufacturing techniques to meet the demands.

The tussling competitive nature of product and service markets calls for the distinctive quality of products and efficiency in service delivery. There is also a need for cheaper products and services. The most certain solution to all of these demands points to the introduction of automation in industries.

What is automation in the industry?

Automation is simply the use of machines and operations which work independently through computer-based software. For an increase in productivity, reduction in the cost of labor, capital, and other material resources, most industries have become technologically driven by implementing automation in industry.

Most automation in the industry uses all sorts of industrial communication devices like **PLCs – Programmable Logic Controllers** and **PACs – Programmable Automatic Controllers** to coordinate most activities that go on in the industry. Therefore, industrial automation is basically about using robots and other technologies to control and coordinate the processes in industries.

An increase in productivity, cost reduction, precision, and flexibility are the basic advantages of using automation. The advancement of the Industrial Revolution saw the rise in mechanization, which led to the production of cheaper goods in high quantity. Before now, the processes of industrial output gave out more speedy and plentiful results, yet there was a need for skilled workers who will prevent the machines from making production errors or waste resources.

But today's automation comes with some control loops, which can either be open control loops or closed control loops. The open control loops accept human input while the closed control loop is automated. With the invention of **Industrial Control Systems** (ICS), the monitoring and control of processes in the industry can be done at the site or remotely off-site. This makes it possible for industries to continually remain operative every day and every hour of the week. Automation has increased efficiency, quality, quantity, and also reduced the propensity for errors. Nonetheless, there are certain limitations or negative aspects of automation. And some of those limitations are retrenchment of workers, high cost of initial purchase and set up, and lack of human ethical evaluations.

Why Is Industrial Automation Today's Trend?

The necessity of automation in industry is innumerable. It improves the quality and time of production. So, a job that might have taken six days to produce a not-too-efficient product is optimized through automation. In other words, automation in the industry makes small input produce bulky and superior results.

Several complex processes can be merged into one automated machine to minimize time, effort, and other tangible and intangible resources. Workers become engaged in higher roles of supervision, and the industry is saved lots of finances reserved for labor.

Once human involvement is reduced, then human-related errors are phased out as well. There's the maintenance of consistency of value, designs, concepts, and creativity. The marginal propensity for error is drastically reduced, and budget for waste management or preemptive errors beneficially dwindles.

You don't necessarily have to wait for a clerk to manually check if you have passed some parameters at the bursary or customer service center when automation can check out everything easily and more effectively. We don't have to worry about the complexities of some processes because automated machines are now very intelligent to handle such for us.

We now have the luxury of carrying out life-threatening situations at any time of the day and night through automated machines. Have you ever imagined how much it will cost bank owners to hire officers who will be dispensing cash all round the clock? But we have our automated teller machines working all day and all night dutifully without complaining and without receiving wages or salaries.

Types of Industrial Automation

Not all automation you see in industries are the same. There are several types of industrial automation. They are:

Fixed Automation

You know it's fixed automation when the pattern of processing operations follows the specifications given by the equipment. Every operation, transaction, or activity in fixed or hard automation is not too difficult to define. When you carefully observe it, you will understand that it comprises the coordination of different operations combined into a piece of equipment; this is what makes it complicated. Fixed automation is not available for everyone's pocket or desire. Its purchasing cost is very high, and it's meant for those who are into mass production. Therefore, before you get fixed automation, you should first ask yourself: is my product in high demand? Can I afford the initial cost? Do I need this now, or should I try other available alternatives? Examples of fixed automation are machine transfer lines, some chemical process instruments, and automatic assembly machines.

Programmable Automation

You know it is programmable automation when it is designed to allow modifications of its program or sequence of operations to suit the various product configurations. The program controls the sequence of operations through the written instructions embedded in it. You'll need this type of automation for products being produced in batches and medium to high volume. The downside of programmable automation is that it's quite difficult to alter the sequence of operations to suit another kind of product. Examples of programmable automation are machines that are controlled by numbers, paper Mills, steel rolling mills, and industrial robots.

Flexible Automation

It is safe to say that everyone loves flexibility, and I think you'll love this automation. A soft or flexible machine is an automated system that can produce a vast amount of products, and you do not have to waste time adjusting it because you want it to suit another product. It is advanced programmable automation. Using a system that is flexible and automated helps you save production time when reprogramming the sequence of the operation or the specifications of the product. You don't have to wait for one batch to finish up before you commence the production of the product. There's no need for separate batches. Examples of flexible automation are CNBC machines, automobiles, driverless cars, and many more.

The advancement of mechanization is industrial automation; it's a step ahead.

Industrial Automation for Quality and Flexibility in Production Process

Automation was initially meant to improve the rate of productivity, and that's why automated devices work 24 hours every day. Another significant reason for the incorporation of automation is to minimize the cost of human labor.

However, today's automation is centered on flexibility and quality in the production process. Let's look at the automobile industry, for instance. The manual installation of pistons into engines was often done with an error rate of 1-1.5%. In our present time of intelligent automation, this same task is done with an error percentage of 0.000001%.

The Advantages of Industrial Automation

There are many benefits of Industrial automation

Lower Operating Cost

Expenses that are accrued from human costs are eliminated by industrial automation. It is usual to have healthcare and sabbatical costs budgeted under the company's expenses. With industrial automation, these expenses are eliminated. Industrial automation does not need benefits such as bonuses, pension, promotion, etc. Even if the cost of the initial payment is high, other expenses like monthly wages of workers are saved up for something else. You may be tempted to enquire: what of the cost of maintenance? But you should understand that maintenance is not actually done very often and it does not break down easily. And when it does, only maintenance and computer engineering would be called to repair it.

Increased Productivity

We have seen companies which recruit a great number of workers in production departments to run three to four shifts round the clock yet there are specific times when the factory would still be shut down for maintenance purpose and vacations. However, industrial automation fills up this gap and keeps the factory running 24 hours every day, and 365 days of the year. This simply increases the entire productivity of a company.

High Quality

With automation, errors associated with human limitations are made minimal. Fortunately, robots do not get stressed out (as far as they are maintained), and they consistently produce quality output.

Increased Accuracy in Information

When data are collated with automation, it gives the analyst the essential production information, accurate data assessment, and reduction in the cost of data research and compilation. The information derived from the data makes you privy to the knowledge you need in making decisions that can help you reduce water and improve data processes.

Improved Safety
These days, humans are not initially sent to very dangerous zones of the world; robots are sent to carry out such tasks. In the factory, employees are not given hazardous tasks that will endanger their lives; robots are used instead.

Disadvantages of Industrial Automation.

Here are some disadvantages of industrial automation:

High Initial Cost
Automation is not cheap. So much initial cost is needed to move from the use of human hands to automatic machines. You will also have to invest a lot in training and retraining workers who will be operating the new advanced machinery.

Fear of Retrenchment
Most workers face this fear. They are frightened that machines are coming to substitute them and take their source of livelihood. But this is not exactly true. Machines may indeed take up the roles of humans in industries, but this will make way for higher productivity, which in turn gives room for more projects to be carried out and supervised by humans.

Loss of Flexibility
It's not easy to edit programmed automation. However, making a proper inquiry from the

vendor exposes you to the right information on how to establish the system.

Industrial automation is here to improve today's world and make every industry of the future more productive, safer, and qualitative in output. It has limitations, although very few that the world can easily afford to pay for.

CHAPTER 5

EXPLORING THE TRENDS

In today's world, automation is no longer a luxury but a necessity because of its significant roles. Its usage helps us to maximize time and makes work more efficient. Intelligent automation has greatly helped in minimizing expenses and increasing resource savings. And now it becomes a trend.

What are the Trends in Intelligent Automation?

PwC has forecasted that by 2030, AI will generate approximately $16 trillion, and 20% of responses from companies surveyed shows that they are already on the move to immensely utilize AI for its benefits this 2020.

There are several types of automation, as new ones are developed at a very fast rate to provide solutions. Companies are changing every day as they try to come up with an automation strategy that can fill up the missing gap of an automation system so that they will have detailed insight on how they can use these tremendous tools to achieve more in business and at home.

Digital Transformation Trends

Not every company has a clear grasp of automation, and those who have are finding it difficult to optimize automation. The first or entry door to automation is Robotic Process Automation or RPA, and this is because its **ROI** is relatively quick.

When companies hear of the trends in intelligent automation, they tend to start planning on how best they can use more automation to achieve the objectives of the business. Nonetheless, **more can be achieved with automation if customers remain the focus of intelligent automation**. Instead of finding ways to determine the use of automation or seeking its improvement, it is more beneficial to businesses when they first identify the needs or wants of customers and then proceed to use intelligent automation to provide solutions.

Unattended and Attended Automation and AI.

Some years ago, users of automation tried automating back-office operations, and their courage gave them success in unattended RPA. These unattended RPAs were scenarios that are a hundred percent automated and, therefore, did not require human input.

Before this development, RPA platforms concentrated on automating front-office operations, which we can also call attended

RPAs. Unlike unattended RPA, they require human interventions, and they can be seen working on desktops.

But as the years evolved to the present, what is trending or common is the complementation of the unattended and attended RPA; that is, the efficient automation of operations which take place in front-office and back-office.

Industries are exploring the integration of AI as well as RPA so that transformative impact can be made in the present and the future. The trend of combining artificial intelligence and intelligent automation is value-oriented.

Governance Gains Focus

Organizations find it difficult to trust technologies that are newly introduced to take over operations. But as we have it today, governmental parastatals are incorporating more automated systems into its workforce. Because of its accuracy, audits are carried out through automation and supervised by humans. Risk evaluation and management in governmental operations are largely done by intelligent automation coupled with artificial intelligence.

Creating Order from Chaos

Change in management, unpredictability, coordination of operations, and auditing take the form of chaos. However, the current trend of things involves using intelligent automation and artificial intelligence to establish order. The secret of most modern successful organizations is that they embrace orderliness by using intelligent automation and artificial intelligence. The trend in intelligent automation is not just about utilizing technology in fulfilling industrial tasks. It's now about how best it can transform work culture and the workforce by ensuring systematic orderliness.

Chief Information Officers Become the Center of Focus

The demand for innovation and greater productivity has made CEOs, CMOs, and CIOs understand that digital transformation and impact is a complex process. Intelligent automation is critically essential in determining the success of their companies.

Organizations now set up board meetings as they strategically plan how they can utilize artificial intelligence and also intelligent automation to their advantage. This situation places CIOs, Chief Information Officers, at the center stage of decision making because their expert contributions have so much to add to the profit of the organization. Their contributions blur the gap between the past and fading systems of operations and the concurrent use of new and impactful technologies that promote values and productivity. It takes professional knowledge and experience to assess new technologies to see how they will fit in the organizational operations and objectives; create an ecosystem that's digitally governed; and provide quality in customer experiences.

Through an interdependent approach that cuts across various departments, CIOs will lead innovative teams in building the best technological model that will satisfy the wants of customers.

Ground Breaking Performance

Firms and users are exploring means in which they can derive greater benefits from intelligent automation. It's now common to see how systems of Intelligent Automation (IA) and Artificial Intelligence (AI) are combined for efficient and excellent performance.

In almost every place of our today's world, intelligent automation is finding its space. Even in information technology, the use of automation is now common and has become popular. Although it was simply made to ease the stress of repetitive tasks, intelligent automation is now a sophisticated technology that enhances advancements and transformation of operations in industries and even in our domestic jobs.

All over you, the effect of intelligent automation is apparent. Because of its efficiency and effectiveness, intelligent automation is now mostly used in the area of information technology for breaking down complex tasks, improving customer experience, and reducing costs.

Automation in Decision Making

The latest innovations in artificial intelligence and automation, which include machine learning and the potential of computers to utilize massive data are very valuable in organizational management. It's now a trend to see intelligent automation and also artificial intelligence involved in the decision-making process of companies.

These technologies give insights beyond the norms of traditional systems that are operated by specific rules. Using intelligent automation in making decisions is trending because it is much faster in pinpointing issues and in the provision of solutions to these issues.

In the operations of information technology (IT), intelligent automation, as well as artificial intelligence, are used increasingly to carry out tasks. Because of the decision-making capacities of these tools, the impact of services is easily monitored, and many resources are preserved.

Test Automation

The deployment of new technologies and services has made test automation a market trend because it is resourceful in promoting value delivery. In information technology, automation now enhances the process from an initial point of "commit and improve" to "test and deployment" by making sure that production is all about guaranteed results.

You, too, will agree that anything which promises to reduce error and ensure productivity will be sought after. That's why test automation is trending. With consistency in test automation, there is a guarantee of product or service performance. So, the industry is satisfied, and the expectations of consumers are met.

Easy Development of Automations

In recent times, intelligent automation has been a trend in almost all facets of life, especially in information technology. And we can resolve that this is because it processes so many complexities and gives outstanding results with or without human input. What's even appealing to everyone is that most automation tools can now be built with low-code processes, and that's what is in fashion right now. It's like today's world of website development and design. Because it's now something anyone can do without even going through the stress of learning so much about coding, almost everybody seems interested in building and owning a website.

With the trend of low-code process, you can see automation tools that are built-in minutes and not just days. You don't have to be a technological guru to build automation. It's now oversimplified and open for everyone, ready to learn the basics and implement.

Keeping Pace with IT Operations

Intelligent automation is a priceless asset in the world of information technology and digital transformation. An example of how automation proves its relevance in information technology is in conducting multiple and complex tasks like the combination of multiple automation workflows across different systems.

Processes that involve the management of network, computerization, and storage tools are now mostly dependent on automation. In management, automation is used to maintain compliance standards.

Most virtual systems that are enhanced to the extent that they bear stark semblance to the real world are operated by intelligent automation.

As intelligent automation, as well as artificial intelligence, permeates society, there is an ease of previous prevalent panic, which was based on the idea that machines are being developed to replace humans. Today, no one is bothered about how automation would take up the roles of humans.

Duties that were once enervating and largely resource-demanding are now mostly automated. Human supervision or judgment is still required, even when automation is becoming more intelligent every day.

CHAPTER 6

ARE THERE PITFALLS?

Everyone wants to solve every problem with intelligent automation. But technological experts do not have this view. A practical overview of the essence of automation shows that automation cannot fix everything, especially office politics, bad infrastructure, lack of training, and failed management. Yet, the position of today's business executives is: automate or die trying. As automation becomes more accessible by almost every business, most employers now want to automate everything.

With the growing simplicity of the business process through the service of automation software, several consultants and companies feel their job is not done until they ensure every inch of the office is enabled or endued with automation.

Intelligent automation, especially **Business Process automation** (BPA), helps to minimize costs and enhance the quality of the result. Human errors are eliminated or reduced because of the accuracy of automation.

Nonetheless, a narrow-minded approach to automation can revert to cause more harm than good. The generalization of automation is a path that leads to failure. Automation cannot provide the solution to all business situations, and if it is generalized in its application, it might lead to more problems.

You may be shocked to know that some tasks are effectively performed without automation software. When these tasks or processes are wrongly handled by automation, it may negatively affect productivity.

Here are the most regular issues you should avoid when utilizing automation.

Overspending

The initial purchase of automation software indeed consumes so much of our hard-earned cash. But you should always remember that automation is meant to cut down costs, save time, and help you maximize profit. So, if you have software that's costlier than it's the rate of productivity or usefulness, then you should know that there's a problem, and you should start making plans to tackle it. The number one index of overspending in automation is when the output indicates that it is a liability and not an asset.

Every company which invests so much in automation does this to make a profit. This is why big companies try to automate so many processes that will yield maximum outcomes. When automation is not well applied, the outcome is usually a painful experience.

Gains which do not Commensurate Investments

As the owner of a business or user of automation, you expect that your investment in intelligent automation will yield better and satisfactory results than what a manual or traditional process would. However, it's quite unfortunate that most companies or users do not sit down to calculate if they are reaping the benefits of their investment. Due to the speed and accuracy of automation, you might assume you're making gains. But are you profiting from your investment, or are you at a loss? You have to watch out and see if the overall performance of the automation is meeting up with your expectation.

As you watch out for your investment, you should also understand that the effect of automation is not usually seen immediately; sometimes, it takes months. Therefore, don't be in haste to draw out conclusions. Even when you realize that you're not making a profit, you shouldn't ditch it yet. Try modifying the setting or referring back to the manual to see where you might be getting it wrong.

It's not enough for automation to be efficient, it should, by all means, be commensurate with the cost of purchase and maintenance.

Allowing Artificiality to Reign over Natural Intelligence

Have you noticed that most times, automated voice service in the customer service sector is both boring and time-consuming? Many customers complain of how it's ineffective because they are always redirected in circles before they get the solution they seek. This is because most automated voice services are not built with artificial intelligence. It's rare to see such automation making adjustments to suit your special requests. But if you study manual voice interaction, you'll see that the originality of the human experience is far better than the artificial automation without natural intelligence.

The key point is that you shouldn't take away the resourcefulness of intelligence because with human intelligence, much can be achieved. It's a fact that automation can perform a complex task countless times without making errors, but some processes need human touch before the final result can be meaningful. Sometimes, automation needs human intervention for proper accountability.

Even customers who the automated software is trying to serve more efficiently may feel ignored when they are never allowed to interact with humans. A perfect example of this is how top websites or social media platforms alienate themselves from their clients through automated responses.

Trying to Automate Everything

Automation software is capable of multitasking repetitive and enervating activities. It can help you handle data management and (online and offline) administration. Complex processes are simplified into workflows, and they become easier to finish off and to manage. But does this mean we should leave everything in the hands of automation? Of course not, It'll be illogical to do such. Trying to leave every task for automation will surely lead to some additional costs, and in the long run, some unexpected complications may come up. You may eventually end up wasting resources that you're trying to preserve. The golden rule of automation is: tasks that require human engagements such as emotions, insights, voice, etc. should be left for humans, and if the tasks can be handled by automation, it's best to leave it for **software bots** or intelligent automation. Automating every task will surely leave your core customers unhappy, and who doesn't know that having unhappy customers is bad for business.

Neglecting Automation

Business and automation of the robotic process usually go on in the background. So, don't be surprised when you find yourself neglecting automation to function without assessment or regular check. If you notice such occurrence, take prompt action to rectify this because so much can go wrong through this pitfall.

Automation runs in a constant process, and it's in your position to make it stop. Laws, technology, and customers can change the processes. In other words, neglecting your automation can lead to the automation of processes that are not current. This means that every output is a waste of resources. Instead of profiting from your investment, you end up counting losses. If your automation is not updated, you probably will find yourself in court settling legal violations; an example is how privacy or data processing laws change in different seasons. Therefore, always test, adjust, and improve your automated processes.

Replacing Humans Totally

One of the major pitfalls of automation is using automation to do the jobs that are best done by humans. Automation can intimidate your workers and make them nervous because they are worried that one day, they will soon be replaced by the new technology. Employees lose morale, and even the most valuable ones may start planning their resignations.

And just as I identified, automation has its limit. There are some human capabilities that automation cannot easily replace. Automation can't think out of the box or solve complex and analytical problems. All of these are human-based problems that can effectively be solved only through the natural faculties of humans.

Every successful enterprise knows that there needs to be a defining path between automation and human intelligence. Automation is neither bold nor awesome. It's best used for repetitive tasks that take place at the back door. It's not too effective in handling frontline tasks. Therefore, a role like a customer representative is best managed by a human.

Thinking Automation Doesn't Need Modifications

Most people think that automation is another software product that you install and begin to operate immediately. However, this is not the situation for automation.

A good way to describe automation is to see it as a platform or as an engine and not a car.

This means that it's like a tool you modify and use for your tasks and processes.

Automation is not easy to set up. It requires time and effort because there'll be changes which it'll need to adapt to.

Finding it Difficult to Differentiate Automation from Artificial Intelligence (AI)

I've come across different articles and books where both technologies were interchanged as the same thing, whereas they are both different, though complementary.

Automation is a tool that does what you want it to do and only excels at its given tasks with its given instructions. Unlike artificial intelligence, automation is not intelligent enough to adjust to unexpected circumstances without human guidance.

That's why automation needs you, or else much of its processes will be incorrect. You'll need to evaluate the final process to correct any found error. Much of the differences between intelligent automation and artificial intelligence has been explained in chapter three of this book.

How to Avoid the Pitfalls of Automations

Automation has helped reduce the expenses of so many organizations, and it has enhanced the efficiency of their processes. Automation is an incredible technology that has been trending and making an impact in our modern world. The reason is that it is an awesome technology that produces fast revenue or return of investment (ROI), **CTO**s everywhere in the world are investing wisely. In media, at conferences, and discussions, so much is being said of its resourcefulness.

So, the pitfalls of automation should not make us ditch the bathwater and the baby. Instead, we should learn how to avoid these pitfalls, and here they are:

Get Everyone Onboard
Keep every personnel informed on the new technology. Educate them on the usefulness of the new technology, its specific roles, limitations, impact, and how it aligns with the company's goal. People are usually afraid of what they can't seem to understand or comprehend, and that is why they fear change. If they are not afraid of the new technology, they might become too excited about it to the degree that they want automation to start doing the most ordinary tasks for them.

Plan to Scale Automation from the Beginning
When you begin to process with automation successfully, you'll realize that it is becoming bigger than you expect. Then you begin to see the necessity of having a support system for such a scenario. It's not possible to automate

single processes and have them remain the way they are forever. When you put the big picture in front of everyone, it becomes easy to adjust to the new development and scale so that it matches the organization's goals. It's not easy to scale automation if you have not planned for it. The situation can be overwhelming. This, therefore, demands that you monitor and continuously optimize the system. Scaling is a beautiful thing when you are equipped with the right strategy and resources to manage it.

Establish Awareness

The personnel in charge of the supervision of intelligent automation need some basic understanding of intelligent automation so that he or she can easily translate everyday processes and instructions for the developers of intelligent automation. This is achievable when there's a budgeted training expenditure, which makes provisions for personnel to be thoroughly educated on the maintenance and resourcefulness of intelligent automation. When the strengths, as well as the weaknesses, are known by management and staff, the walk to organizational growth has begun.

Knot the Ties of Information Technology and Automation

Automation serves you well when it has a healthy relationship with IT. Automation goes via the user interface and imitates the processes of a user. Just a little change will woefully affect the functionality of the automation and might even crash the entire process. This can be avoided when open communication exists between the automated device and the application managers. The automated system will need some form of connection with your company's network, and it will require so much effort to set up an enabling framework that can accommodate the ROE – Robotic Operating Environment.

Ensure the Tasks Need Automation

Most companies want to believe automation is the master key that must unlock or handle every problem when, in reality, this is not how it is. Automation should not be given to every process. The process needs to be something that can be predicted because anything or situation that's beyond the rules of the program may likely be regarded as an exception that cannot be processed. Even Bill Gates affirmed that automation would not solve inefficiency because automating a process that's already inefficient will only maximize the inefficiency. Before an automated device is used for a particular purpose, the processes ought to be theoretically and practically analyzed before it's bought or implemented.

100% Shouldn't be Given to Automation

Just as I have said, some processes should not be handled by automation. Even if the job is meant for automation, sometimes, 100% of the job shouldn't be left for automation. In other words, it's best if you automate some aspects of a task. For instance, if 80% of a task is highly predictable, and the rest is unpredictable, automation is the best deal for the former, while human intervention is best for the latter.

Intelligent automation is very beneficial if it is implemented in solving problems, especially those that are predictable. Although automation comes with some major pitfalls, there are strategies in place to tackle most of them. And when we avoid these pitfalls, we shouldn't just be contented with the advancements we see today. We should be planning for the future because the future really belongs to the ones who have planned for it.

In our next chapter, we'll be having a take-off into the future of automation. Much insight will be revealed for your benefit. So, get comfortable, and let's pay the future of automation a visit.

CHAPTER 7

THE FUTURE OF INTELLIGENT AUTOMATION

"What will the future look like?"

"In the workplace, will intelligent machines substitute humans?"

"Are the sci-fi movies on technology real?"

These are contemporary questions we joggle in our minds. You want to know your place in the future and then see how you can leverage this knowledge for a better reward in the future.

The Future Includes You

The impact of intelligent automation cannot be argued. It's there for everyone to see. Yet we have seen that irrespective of how much machine learning and advancement is given to automation, it will always rely on humans for some specific responsibilities. It's more likely that intelligent automation will be assigned tasks that are based on rules and routines such as data entry or processing of orders, while humans will intervene in solving tasks that demand human sociological and emotional intelligence.

It's not easy to develop automation that can imitate the peculiar traits of humans. Traits such as creativity, intuition, and empathetic communication cannot be easily built into machines.

Therefore, humans will not be replaced by automation. Instead, there'll be a restructuring of the workforce that will be favorable to employees, employers, consumers, and automation. Employees will be trained to perform tasks that demand human creativity and uniqueness. Automation will deliver roles that are not too socially or emotionally demanding.

The Place of Automation in the Future

Then what processes the best suit the functions of automation in the future?

Intelligent automation is excellent at carrying out manual tasks which have defined pattern of operation or repetitive. When we say automation is excellent in carrying out this type of task, we simply mean it can perform these tasks over and over again without making errors.

Although it's preferable to have automation take care of activities which are template-driven and can be carried out under the performance of already processed instructions, automation in the future will be designed to be highly independent using more of deep learning and new inventions.

Unlike now where automation is only given tasks which have been assessed to be of high volume and frequency, low rate of the variability of the outcome, distinct type of input, intelligent automation will be designed to work on novelty and a paradigmatic shift towards performing unpredictable tasks.

More on the Future

Just like CiGen client, which is an Australian FMCG food company, delivery confirmations will be automated in Enterprise Resource Planning (ERP). This system allows intelligent automation to read, extract, and rename PDF files enclosed in dockets to upload the documents. Through **NetSuite cloud ERP**, it identifies if the sales order corresponds. If there's any incidence of unmatched delivery, it makes a report.

This reduces processing time by 95%, and the completion of process time runs at 100%; that zero-chance of errors. More of this kind of development in the future means there'll be faster delivery and time of remittance, quick resolution of conflicting issues, and an update on delivery status.

In the utility sector, the database of asset management can and should be automated to ensure that the stringent regulatory compliance **SLA's** can be met. Intelligent automation will make data entry accurate, and processing time becomes faster.

CONCLUSION

The future of our world is one that's overwhelmed with intelligent automation. In pharmacy, business, education, homes, food, movie, technology, and virtually every sector of the world, intelligent automation will be the passport to efficient and productive outcomes.

ABOUT THE AUTHOR

Elijah Falode is a Top-Rated Content Writer on Upwork and a Professional Content Development Consultant with over ten years of experience in educating people about business and emerging technologies.

In the past five years, he has collaborated and worked with thought leaders and industry experts as a content development consultant.

He has researched, and ghostwritten content on emerging technologies- AI, Voice economy, Big Data, Connected Cloud, RPA, IoT and Cybersecurity, and He is fully aware of their capabilities and risks.

One of the fears He doused in this book is that of people being afraid of losing their job if and when every process is automated. Of course, there will be a job loss, but you can embrace the future of technologies when you improve your learning and skills on emerging technologies.

"Only those who are prepared will be relevant in a few years to come."

REFERENCES

1. S. Russell and P. Norvig, "A Modern, Agent-Oriented Approach to Introductory Artificial Intelligence" ACM SIGART Bulletin April 1995 https://doi.org/10.1145/201977.201989

www.ingramcontent.com/pod-product-compliance
Lightning Source LLC
Chambersburg PA
CBHW051539240526
45465CB00027B/726